Wood and Paper

Developed at
Lawrence Hall of Science
University of California at Berkeley

Published and Distributed by **Delta Education**

© 2007 by The Regents of the University of California. All rights reserved. No part of this book may be reproduced or transmitted in any form or by any means, electronic or mechanical, including photocopying or recording, or by any information storage and retrieval system, without permission in writing from the publisher.

542-0047
ISBN-10: 1-59821-023-8   ISBN-13: 978-1-59821-023-1
5 6 7 8 9 10 QUE 13 12 11 10 09 08

Table of Contents

The Story of a Chair 3
Are You a Scientist? 9
The Story of a Box 13
Land, Air, and Water 19
I Am Wood 24
Glossary 25
Photo Descriptions 27
Index 28

The Story of a Chair

Here is a chair you might sit on in a park.
Do you know how it was made?

A chair begins as a **tree** in the **forest**.

A lumberjack cuts the trees in the forest.

A truck driver takes the trees to the sawmill.
The trees are cut into boards at the sawmill.

The boards go to the lumberyard.
The woodworker buys **wood** at the lumberyard.

The woodworker makes
the chair from wood.
And it's put in the park for you
to sit on!

Are You a Scientist?
Scientists ask questions.

Scientists look at things to find out about them.
They touch, smell, and listen to things, too.

Scientists **compare** things. They look at how things are the same and different.

Scientists talk about what they find out.

Are you a scientist?
You are!

The Story of a Box

This box is strong.
But it is made of **paper.**
Do you know how a box is made?

A box starts in the forest as a tree.
First, the trees are cut.
Then, the trees go to a paper mill.

The trees are chopped to make **sawdust.**
Sawdust is soaked in water.
This makes **pulp.**

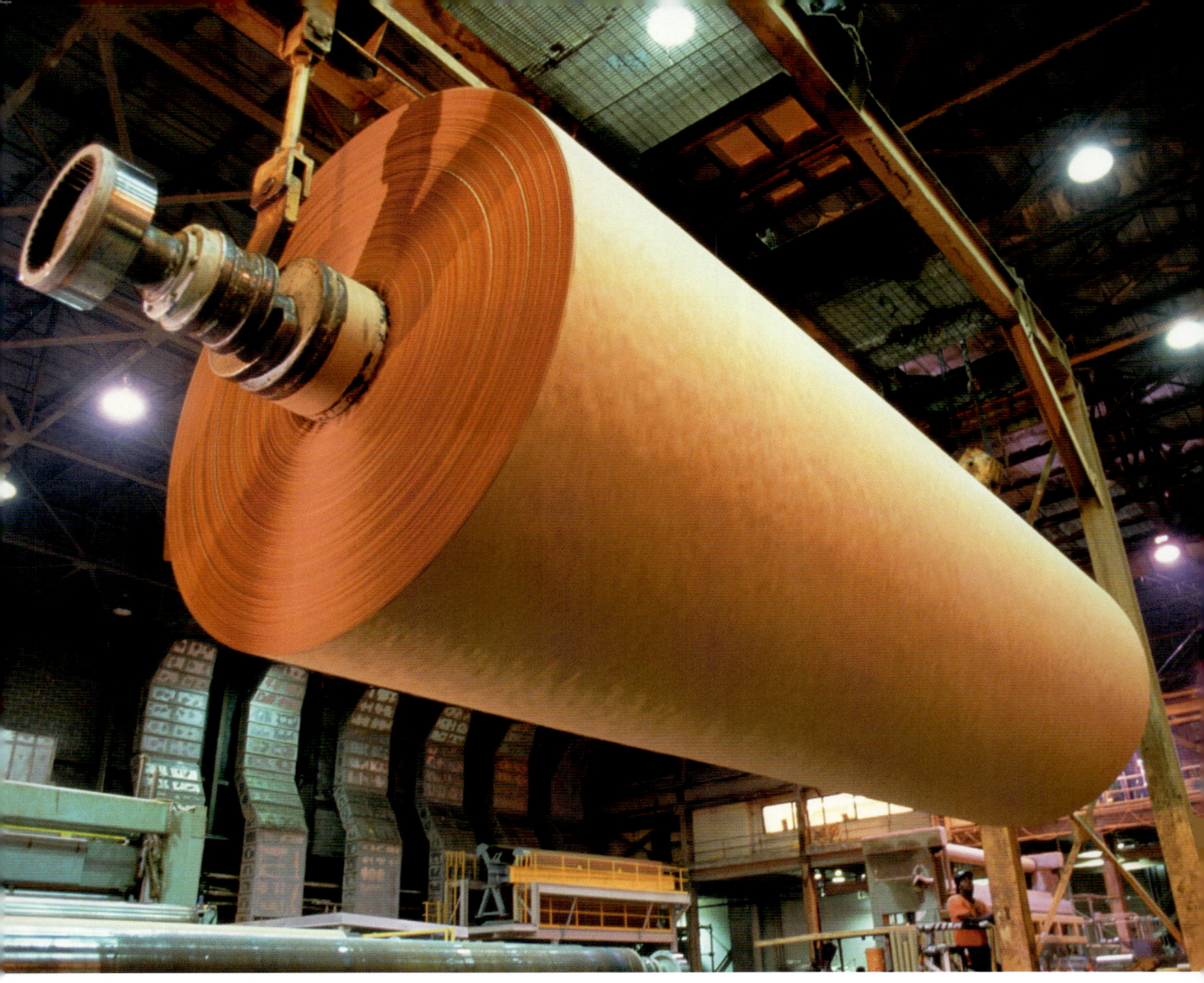

The pulp is flattened and dried to make paper.
The paper is put on a roll.
This roll of paper can be used to make cardboard.

The paper is layered and glued together.
This makes cardboard.
The cardboard is cut and folded.

Now the box is ready to use!

Land, Air, and Water

Do you ever **recycle** paper? If you do, you are helping to save a tree.

We use trees for many things.
Trees are important to us.
Other things are important
to us, too.

Land, air, and **water** are important.
We need good land for growing **food.**

We need good water to drink.

We need good air to breathe. How can we take care of our land, air, and water?

Glossary

air a mixture of gases you can't see that is all around you

compare to look at how things are the same and different

food something plants and animals need to live

forest a natural area where trees grow

land the solid surface of Earth covered with soil and rock

paper something you write or print on. Paper is made from trees.

pulp a liquid made from wood. Pulp is used to make paper.

recycle to use again

sawdust the small bits made when a saw cuts wood

scientist a person who studies the natural world

tree a plant with a wood stem, roots, and branches with leaves

water a resource on Earth that can be changed from a liquid to a solid

wood what is under the bark of trees. Tree trunks, branches, and stems are made of wood.

Photo Descriptions

Cover: A wooden house frame with blueprints on the floor

Page 1: Making papier-mâché bowls

Page 2: A man carving a totem with children

Pages 3 and 8 (right): A white park bench with flowers

Page 4: A young pine tree with trunk of a large tree in the background

Page 5: A lumberjack cutting down a tree

Page 6 (left): A truck with logs leaving the forest; (right): logs at a sawmill

Page 7: A forklift lifting boards at a lumberyard

Page 8 (left): A woodworker making a chair

Page 9: Kindergarten students observing and asking questions about sawdust and wood shavings in water

Page 10: A kindergarten student observing different wood

Page 11: Kindergarten students comparing how many paper clips it took to sink a piece of plywood versus particleboard

Page 12: A kindergarten teacher conducting the Wrapping-up for Investigation 2 of the FOSS Wood and Paper Module

Page 13: A cardboard box on a pallet in a warehouse

Page 14: Logs being cut with heavy machinery into shorter lengths for milling at a sawmill

Page 15: Wood pulp at a newspaper mill

Page 16: Brown paper on a roller in a paper factory

Page 17 (top): A worker scoring cardboard sheets to make boxes; (bottom): a cardboard box showing three different layers of paper

Page 18: Cardboard boxes moving on a conveyer belt

Page 19 (top): People recycling at a neighborhood recycling center; (bottom): a recyclable grocery bag surrounded by produce

Page 20: Bridger-Teton National Forest and Gros Venture River in Wyoming

Page 21: Children working in a garden

Page 22: A child drinking from a water fountain

Page 23: Two girls holding onto a signpost while the wind blows

Page 24: A girl in the forest (background)

Index

Air, pp. 21, 23

Box, pp. 13–14, 18

Cardboard, pp. 16–17

Chair, pp. 3–4, 8

Compare, p. 11

Food, p. 21

Forest, pp. 4–5, 14

Land, pp. 21, 23

Lumberjack, p. 5

Lumberyard, p. 7

Paper, pp. 13, 16–17, 19

  Process of making, pp. 13–18

Paper mill, p. 14

Park, pp. 3, 8

Pulp, pp. 15–16

Recycle, p. 19

Resources (concept of), pp. 19–23, 24

Sawdust, p. 15

Sawmill, p. 6

Scientist, pp. 9–12

Senses (concept of), pp. 10–11

  Sight, pp. 10–11

  Smell, p. 10

  Sound, p. 10

  Touch, p. 10

Tree, pp. 4–6, 14–15, 19–20

Water, pp. 15, 21–23

Wood, pp. 7–8, 24

Woodworker, pp. 7–8